Anatomy of fuse

Guidelines On How To Use A Fuse Properly

Dr Walt wade

Contents

Chapter1 3

   Introduction to anatomy of fuse 3

Chapter2 9

   Construction and working of a fuse 9

chapter3 21

   characteristics of a fuse 21

chapter4 27

   how to select a proper rating size of a fuse 27

The end 33

Chapter 1

Introduction to anatomy of fuse

Fuse anatomy refers to the intricate and detailed structure of a device that consists of a connecting element and a protective element, commonly used in electrical systems. It serves as a safety mechanism that prevents electrical circuits from overloading or short-circuiting, which can lead to serious consequences like fires or damage to electrical equipment. Fuses have been used for centuries, with early civilizations using strips of tin or copper to provide protection against electrical hazards. The anatomy of a fuse is essential to understand as it plays a crucial role in its functionality. It comprises several parts, including the body, cap, element, and markings, each with a specific function to ensure safe and efficient operation. In this article, we will delve into the anatomy of a fuse and explore the various components and their functions. One of the primary components of a fuse is its body, which is made of a non-conducting material such as glass, ceramic, or plastic. The body not only provides mechanical support but also serves as insulation between the two terminals of the fuse. It also protects the fuse element from external environmental factors that can cause damage or corrosion. Another crucial component of a fuse is the cap, also known as the end cap or ferrule, which connects the fuse element to the body. The cap is usually made of metal, such as brass or copper, and is responsible for conducting the electrical current from the circuit to the fuse element. It also helps to secure the element in place while maintaining a strong electrical connection. The fuse element is the heart of the fuse and is responsible for interrupting the electrical circuit when an overload or short-circuit occurs. It is typically made of a thin strip of metal, most commonly copper or silver, that has a lower melting point than the circuit's maximum current. When the current exceeds the fuse's capacity, the element heats up, causing it to melt and break the circuit, thus preventing any further damage. Apart from the physical components, the anatomy of a fuse also includes markings that provide crucial information about the fuse's specifications and ratings. These markings can include the manufacturer's name, voltage and current ratings, and the type and size of the fuse. It is essential to understand these markings to ensure the right fuse is used for the specific electrical system. Fuses can be categorized into two main types – cartridge fuses and plug fuses. Cartridge fuses are commonly used in larger electrical systems, such as industrial and commercial settings, while plug fuses are usually found in residential settings. They differ in their shape, size, and method of installation, but their anatomy follows the same basic principles. The cartridge fuse has a cylindrical body with two metal end caps. It is designed to be inserted into a fuse holder or block, which keeps the fuse secure and provides a connection to the electrical circuit. Cartridge fuses can be further divided into two subtypes – ferrule fuses and blade type fuses. Ferrule fuses have cylindrical end caps, while blade type fuses have flat blades that are inserted into corresponding slots in the fuse holder. On the other hand, plug fuses have a larger and more block-like shape, with metal screw-in caps on each end. They are inserted into a fuse box or panel and have a filament that connects the two end caps. They are commonly used in household appliances and lighting fixtures. In addition to the type of fuse, there are also different classes of fuses, which indicate their speed and response time when an overload occurs. These classes include fast-acting, slow-blow, and time-delay fuses, each with its own unique characteristics and applications. The anatomy of a fuse is not limited to its physical components but also extends to its behaviour and performance under different conditions. For instance, fuses can have a range of voltage and current ratings, which determine their maximum capacity and the level of protection they provide. It is crucial to choose the right fuse with the appropriate ratings to ensure proper functioning and avoid potential hazards.

Chapter 2

Construction and working of a fuse

The construction of a fuse is fairly simple. It consists of a fuse element, a casing, and two end caps. The fuse element is the most important part of the fuse and is made of a low melting point metal such as copper, silver or aluminum. The casing is usually made of glass or ceramic, which makes the fuse transparent so that the condition of the fuse can be easily seen. The end caps are made of conductive metal and are used to connect the fuse to the electrical circuit. The working principle of a fuse is based on the heating effect of electrical current. When an electrical circuit is closed, current flows through the circuit and through the fuse as well. The fuse element is designed to have a lower melting point than the conductors in the circuit. This means that if the current passing through the fuse exceeds a certain level, the fuse element will get heated up and eventually melt, breaking the circuit and stopping the flow of current. This process happens very quickly, usually within milliseconds, and helps to prevent overheating and damage to the electrical system. To understand the working of a fuse in more detail, let's look at the different types of fuses and their construction and working: 1. Low voltage fuses Low voltage fuses are the most commonly used fuses and are used in household and industrial circuits with voltages up to 1000V. These fuses have a cylindrical shape and are mostly made of glass or ceramic. The two end caps are made of brass or copper, which acts as a conductor. The fuse element, which is the most important part of the fuse, is usually made of a thin strip of copper or silver. This thin strip has a predetermined thickness and length, depending on the amount of current it is designed to carry. When the current exceeds this predetermined value, the fuse element gets heated up due to its resistance to the flow of current and eventually melts, breaking the circuit. 2. High voltage fuses High voltage fuses are used in circuits with voltages above 1000V. These fuses are larger in size and are usually filled with a special type of oil or gas that acts as an insulator. The fuse element in high voltage fuses is made of a longer and thicker strip of metal compared to low voltage fuses. This is because a higher voltage requires more current to cause the fuse element to melt. 3. Fast-acting fuses Fast-acting fuses are designed to blow quickly when there is a sudden surge of current. These fuses are commonly used in circuits that are sensitive to changes in current, such as those found in electronic equipment. The fuse element in these fuses is made of a thinner strip of metal, which allows it to melt faster and prevent any damage to the equipment. 4. Slow-blow fuses Slow-blow fuses, also known as time-delay fuses, are designed to withstand temporary overloads without blowing out. These fuses are commonly used in circuits that have motors or other equipment that require high starting currents. The fuse element in these fuses is thicker and longer, which allows them to withstand a higher current for a longer period of time before melting. Apart from these types of fuses, there are also miniature fuses, cartridge fuses, and blade fuses, each with a specific construction and working principle. However, all of these fuses work on the same fundamental principle of breaking the circuit when the current exceeds a certain level. In addition to their construction and working, it is also important to understand the different ratings and markings on a fuse. Fuses are rated in terms of their current-carrying capacity and voltage rating. The current rating is the maximum amount of current that the fuse can carry without blowing out. The voltage rating is the highest voltage that the fuse can withstand before breaking down. This information is usually marked on the casing of the fuse, along with other important details such as the type and size of fuse.

construction of a fuse

Construction: The construction of a fuse is quite simple and it generally consists of a thin strip of metal that is placed within a cylindrical body made of ceramic, glass, or plastic. The most commonly used metal for fuse elements is lead because it is easy to shape, provides low resistance and has a low melting point. Other metals such as copper, aluminum, and silver are also used, depending on their specific properties and intended use of the fuse. The cylindrical body of the fuse usually contains two metal end caps, which are connected to the circuit. The end caps are made of materials that provide good electrical conductivity, such as copper or brass. These end caps are then attached to the fuse element, and the entire assembly is sealed with a filling material such as sand or quartz. Working: The working principle of a fuse is based on the concept of electrical resistance and the property of metals to melt at high temperatures. In a circuit, the fuse is connected in series and placed on the path of the current flow. When the electricity flows through the fuse, the fuse element offers resistance to the flow of current. This creates heat, which causes the fuse element to heat up and eventually melt if the current exceeds the rated value. As the current reaches the maximum allowable limit, the heat generated becomes sufficient to melt the metal strip, creating a gap in the circuit. This gap prevents further flow of current and effectively breaks the circuit. Once the fuse element is melted, the circuit becomes open and interrupts the power supply, providing protection to the circuit and the connected devices. Types of Fuses: Fuses are available in various types, each designed to cater to specific applications and requirements. The most commonly used fuses are as follows: 1. Cartridge Fuses: These are the most widely used type of fuses and consist of a cylindrical body that contains the fuse element. They come in various shapes and sizes, ranging from small glass tubes to larger ceramic tubes, depending on their voltage and current rating. 2. Blade Fuses: Also known as automotive fuses, these are flat-shaped fuses and are commonly used in vehicles. They have a plastic housing with metal blades on the sides, which insert into the fuse box to make the connection. 3. Plug Fuses: These fuses have a metal base with a threaded design that can be screwed into a socket.

They are mostly used in residential and commercial buildings to protect household appliances and devices. 4. Resettable Fuses: Unlike traditional fuses, these fuses have the ability to recover from a fault and can be reused. They consist of a special polymer-based material that heats up and expands when there is an abnormally high current, causing the circuit to break. Once the fault is eliminated, the polymer cools down and hardens, restoring the circuit. Importance of Fuses: Fuses are an essential component in every electrical system because they provide protection against potential hazards. Some of the major reasons why fuses are important are as follows: 1. Safety: Fuses play a vital role in ensuring safety and preventing hazards caused by the flow of excessive current. By disconnecting the power supply when there is a surge in current, fuses prevent fire accidents, electrocution, and damage to property and equipment. 2. Cost-Efficient: Fuses are relatively inexpensive compared to other protective devices such as circuit breakers. This makes them a cost-effective solution for overcurrent protection, especially in small-scale applications and devices. 3. Easy to Replace: Fuses are easy to replace, and in most cases, it only takes a few minutes to get the circuit up and running again. This is particularly useful in situations where the fault is caused by external factors and can be easily rectified. 4. Fast Response Time: Fuses have a fast response time, which means they can quickly break the circuit when there is an overflow of current. This prevents any further damage to the circuit or connected devices.

chapter3

characteristics of a fuse

1. Fuse Materials The most common materials used to make fuses are metals such as copper, silver, aluminum, zinc, and lead. These metals have low melting points and high electrical conductivity. The metal strip or wire forms the main part of the fuse and determines its current rating. In some cases, ceramic, glass, or other non-conductive materials are also used to make fuses. 2. Current Rating The current rating of a fuse is the maximum amount of current that can flow through it without melting the fuse element. It is usually indicated in amps (A) and is determined by the size and material of the fuse element. Fuses with higher current ratings are used for

heavy-duty electrical equipment, while those with lower ratings are suitable for smaller devices. 3. Voltage Rating A fuse's voltage rating is the maximum voltage that the fuse can withstand without breaking down. It is usually specified in volts (V) and depends on the insulation material and thickness of the fuse. A higher voltage rating ensures that the fuse will not fail due to voltage surges or transients. 4. Time-Current Characteristics Fuses have different time-current characteristics, which describe how long it takes for the fuse to open the circuit when overloaded. Fast-acting or quick blow fuses have a short opening time, typically less than a second, and are used for sensitive and expensive equipment. On the other hand, slow-blow or time-delay fuses have a longer opening time, usually a few seconds, and are used for inductive loads, such as motors and transformers, which have high inrush currents. 5. Breaking Capacity The breaking capacity of a fuse is the maximum amount of short circuit current that it can safely interrupt without damaging itself. It is generally expressed in kA (kiloamps) and is an important consideration when choosing a fuse for an electrical system. A fuse with insufficient breaking capacity can cause an electrical fire or explosions, endangering the safety of personnel and damaging the equipment. 6. Fuse Enclosure In high voltage and high power applications, fuses are usually enclosed in a container or box to protect them from environmental factors and reduce the risk of electric shock. The fuse enclosure also prevents the sparks and hot gases released during a fuse's operation from causing any damage or harm. 7. Temperature Sensitivity The temperature sensitivity of a fuse is its ability to withstand high temperatures before it melts. When a high current flows through the fuse, the fuse element heats up and gradually melts. The temperature at which the element melts is affected by several factors, such as the fuse material, insulation materials, and surroundings. In high ambient temperature environments, fuses with a lower temperature sensitivity are preferred to ensure their proper operation. 8.

Reliability Fuses are known for their reliability, mainly because they have no moving parts. Unlike other protective devices such as circuit breakers, fuses do not require any mechanical mechanism to operate. This eliminates any possibility of mechanical failure and ensures that the fuse will operate when needed. 9. Reusability One of the main drawbacks of using fuses is that they are not reusable. Once a fuse has blown, it needs to be replaced with a new one. In some cases, resetting a tripped circuit breaker is more convenient than replacing a fuse. However, this reusability feature of circuit breakers also makes them more expensive. 10. Inexpensive Fuses are one of the most economical options for electrical protection, making them a popular choice for low-cost applications. They can be mass-produced using relatively simple manufacturing methods, and their basic design has not changed significantly over the years. This also makes them readily available and easy to replace if needed.

chapter4

how to select a proper rating size of a fuse

Understanding the Basics of Fuses Before we dive into selecting the right fuse size, it is essential to understand some basic concepts of a fuse. A fuse is rated in terms of its current-carrying capacity, voltage rating, and interrupting capacity. The current-carrying capacity is the maximum amount of current that a fuse can safely pass without blowing. The voltage rating is the maximum voltage that the fuse can withstand. The interrupting capacity is the maximum fault current that the fuse can interrupt without damage. Factors to Consider While Selecting Fuse Size Several factors need to be considered while selecting the right fuse size. 1. Equipment's Operating Current: The first and foremost thing to consider while selecting a fuse size is the equipment's operating current. You can determine the operating current by referring to the equipment's specifications or consulting with the manufacturer. The fuse must have a current rating that is equal to or slightly higher than the equipment's operating current. Choosing a fuse with a lower current rating can lead to frequent blowing, while a higher current rating will not provide adequate protection. 2. Type of Load: Another crucial factor to consider is the type of load your equipment is powering. There are two types of loads - resistive and inductive. Resistive loads have a constant current, while inductive loads have a higher starting current than their operating current. It is crucial to consider the type of load while selecting a fuse size as inductive loads can cause the fuse to blow even when the current is within its rated value. 3. Ambient Temperature: The operating temperature of the equipment also needs to be considered while selecting a fuse size. Fuses have a temperature derating curve, which indicates the reduction in its current-carrying capacity with an increase in temperature. For example, a fuse rated for 10A at 25°C may only be able to carry 9A at 50°C. It is essential to choose a fuse size that can withstand the operating temperature of the equipment to ensure proper protection. 4. Short-

Circuit Current: The fuse must be able to handle the maximum short-circuit current that the equipment or circuit can produce. If the short-circuit current exceeds the fuse's interrupting capacity, it can lead to equipment damage or fire hazards. Therefore, it is essential to determine the maximum short-circuit current of the equipment and choose a fuse with a higher interrupting capacity.

5. Temperature Rise: Excess heat generated by the fuse can cause damage to the equipment and surrounding components. When selecting a fuse size, it is crucial to consider the temperature rise of the fuse. The amount of heat produced by a fuse depends on its material, design, and current rating. It is recommended to choose a fuse with a

lower temperature rise to ensure the safety of the equipment. 6. Type of Fuse: There are different types of fuses, such as fast-acting, time-lag, and ultra-rapid, each designed for specific applications. Fast-acting fuses react quickly to overcurrent and are used for sensitive equipment. Time-lag fuses have a delay circuit to tolerate small inrush currents and are suitable for inductive loads. Ultra-rapid fuses are used for high-speed applications such as motor protection. It is important to select a fuse type that is suitable for your equipment's application. 7. Redundancy: If your equipment is critical and cannot afford any downtime, you may consider using redundant fuses. Redundancy is achieved by using multiple fuses in parallel, each having its current-carrying capacity. In case one fuse blows, the other fuse will provide protection to the equipment, preventing any downtime. When using redundant fuses, it is crucial to ensure that the total rated current of all fuses does not exceed the equipment's operating current.

The end